洪德成 设计作品 生活 艺术
INTERIOR DESIGN LIFE & ART

洪德成 著

大连理工大学出版社

图书在版编目 (CIP) 数据

洪德成设计作品：生活　艺术 / 洪德成著. —大连：大连理工大学出版社, 2014.1
 ISBN 978-7-5611-8292-5

Ⅰ. ①洪… Ⅱ. ①洪… Ⅲ. ①室内装饰设计 – 作品集 – 中国 – 现代②艺术 – 作品综合集 – 中国 – 现代 Ⅳ. ① TU238 ② J121

中国版本图书馆 CIP 数据核字 (2013) 第 244074 号

出版发行：大连理工大学出版社
　　　　　（地址：大连市软件园路 80 号　邮编：116023）
印　　刷：利丰雅高印刷（深圳）有限公司
幅面尺寸：280mm×280mm
印　张：33
插　页：4
出版时间：2014 年 1 月第 1 版
印刷时间：2014 年 1 月第 1 次印刷
策划编辑：袁　斌　刘　蓉
责任编辑：刘　蓉
责任校对：王　琴
装帧设计：刘竞华

ISBN 978-7-5611-8292-5
定　　价：428.00 元

电话：0411-84708842
传真：0411-84701466
邮购：0411-84703636
E-mail:designbookdutp@gmail.com
URL:http://www.dutp.cn

如有质量问题请联系出版中心：（0411）84709246　84709043

Foreword | 自序

GRATITUDE, EMOTION, FEEL AND INSIGHT
感恩、感情、感觉和感悟

Gratitude

I've struck on an idea for stopping and summarizing myself for the last 56 years experience. Review the past and think of what have happened during the period, all of these can be reflected in the book. When I was child, I was influenced by my family artistic atmosphere, and to be a student of Decorating Department of Guangzhou Academy of Fine Arts and dreamed to be a painter, then came to Shenzhen, the Special Economic Zone during the Chinese reform and opening up. It has been 30 years since I was engaged in the design art industry, and I never change my mind to continue my design art road.

Mother, the first person I should be grateful to, and her love is great. Mother was kind, diligent and wise, the typical Chinese female image, and she was the staunch supporter of my career pursuit. Last year, my 90-year-old mother went to the Western Paradise World serenely, as a son, I am never ready to accept the fact, and my life is occupied by the guilty of not spending enough time to accompany her. When I calm down, stop working and come back to life, mother often appears in my dreams. It seems that she never leaves me.

I would also be deeply grateful to the people who loved me, helped me when I had difficulty, these people make me happy. Thanks for my first company—Hong Kong De Sheng Design & Egineering Company, I was fortunate to participate in the interior design of International Trade Building, known as the tallest building at that time in China. Thanks for those friends home and abroad, colleagues, and people supported me during my artistic dream journey, thanks again, my design works have footprints of all of you. I will never forget it.

Emotion

Thanks for my wife giving births to my daughter and son, taking care of them. And now, my daughter undertakes the arts career after graduating from college, my son is educated in University in the United States. My wife devotes much to the family and our ideals, all success in my career is owing to her support, and we are accompanied by each other in the rest of our lives.

Feel

The sense of design art originates from the innate and acquired disposition. I was born and grew up in an artistic family; my parents moved back to the mainland from Hong Kong, China in 1940s. I remembered when I was 8 years old, my elder brother, Dezhong was enrolled in the Technology Academy of Fine Arts, he often painted till very late in the night, and I always stayed with him and all I could do was to help him to wash pen and do some cleaning. Once I got a slap as the pen was dirty, and this slap started my design art career and let it become my lifetime career.

Insight

I choose the design art, no one forced me, because I like it by my heart, having no connection with the commercial benefits. When I look back at my 30-year design career, I find that I still prefer to work in project spot with our design team. As the old saying goes, enough is as good as a feast. I have had a new life attitude since several years ago; I might embrace and enjoy nature with my sense of art. So I started to go to different places to find the most simple and real art of living by my digital single-lens reflex camera. Meanwhile I do some free and romantic painting in my leisure time by my brushes, to get back the feeling of studying painting and sketching in my childhood and enjoy a truly relaxing artistic life. Personally, I hope that the publication of the book is a channel of exchanging the design idea and sharing the joy of life with our customers, designers and artists.

感恩

在我人生经历的这第 56 个年头，特别想停下来总结一下自己。通过这本书，回顾一路走来的点点滴滴。从幼年家庭艺术氛围的熏陶，到就读广州美院装潢系梦想成为一名画家，再到投入中国改革开放深圳经济特区的建设，坚持至今，已是三十载设计艺术之路，再苦再累都没动过改行的心。

母亲是人世间我第一要感恩的人，母亲的爱是伟大的，我在母亲身上看到了一个善良、勤奋、智慧的中国女性形象，母亲也是我事业追求的坚定拥护者。去年，90岁的母亲安详地去了西方极乐世界，作为儿子这是永远无法做好准备的事，我一生难以释怀的歉疚就是没能停下来，多陪陪老人家。当我静下来，从工作回到生活中，母亲的身影时常出现在我的梦中，她似乎从不曾离开过。

我还要感恩所有爱过我、帮过我的人，在我有困难的时候给过一个小方便的人，正因为有了他们，我才快乐到今天。感谢最早聘请我的香港得盛设计工程公司，让我有幸参与到当时被称为中国最高楼的国贸大厦内部装饰项目中。感谢在我实现设计艺术梦想的旅程中，那些和我一起共事过的中外朋友、同事们，在这里说一声衷心的感谢，在我今天的设计成果里有你们每个人的足迹，令我终生难忘。

感情

感恩我的太太，不但为我生下女儿和儿子，而且都将他们培养成人，女儿大学毕业后从事艺术工作，儿子在美国上大学。我的太太为家庭、为我们共同的理想事业付出了很多很多，有她才有我今天事业的成功，有她才有余生的真情相伴。

感觉

对设计艺术的感觉有先天的，也有后天的。我出生并成长在一个艺术世家，父母于20世纪40年代从中国香港迁回内地。记得我8岁的时候，大哥德忠考取了工艺美院，经常关上门一画就到深夜，我时常陪伴左右，主要的工作是帮他洗笔和打扫卫生。有一次因为笔没有洗干净被狠狠打了一巴掌，正是这一巴掌让我走上了设计艺术这条路，并成为了我的终生事业。

感悟

玩了一辈子设计艺术，没人强迫我。我从心底里喜欢艺术，也没想过能挣多少钱。30年的职业生涯一路走来，蓦然回首，我依然还带着设计团队战斗在第一线。古语有云：知足常乐。因此，我在几年前开始有了一个新的生活态度，用我的艺术感觉去怀抱大自然，享受大自然。我拿起了单反数码相机去到世界不同的角落，寻找最质朴、最真实的生活艺术。同时我也重拾画笔，在闲暇的时候沉浸在自由浪漫的绘画创作中，回归到青少年时代学画、写生的感觉中去，享受真正属于自己的随心放松的艺术创作生活。希望通过本书的出版，与我们的客户和同行设计师、艺术家们共同交流，分享人生的快乐。

Dickson Hung 洪德成

Chairman and Design Director of Dickson Hung Associates Design Consultants(HK)Co.,Limited, Founder and Design Director of DHA International Design Alliance Executive

MDM of Politecnico di Milano

Member of Hong Kong IDA Interior Design Association

Councilor of Asia-Pacific Hotel Design Association

洪德成设计顾问（香港）有限公司及意大利DHA国际设计顾问机构创办人／设计总监

意大利米兰理工学院设计管理硕士

香港IDA室内协会会员

亚太酒店设计协会理事

CONTENTS
目录

10
RICH'S CITY CLUB
君华新城会所

36
KINGKEY TIMEMARK SALES CENTER
京基滨河时代售楼中心

46
KINGKEY TIMEMARK SHOW FLAT A
京基滨河时代样板房一

61
KINGKEY TIMEMARK SHOW FLAT B
京基滨河时代样板房二

66
CINEASTE FOUR SEASON GARDEN CLUB
益田影人四季花园会所

72
CINEASTE FOUR SEASON GARDEN SHOW FLAT A
影人四季花园样板房一

81
CINEASTE FOUR SEASON GARDEN SHOW FLAT B
影人四季花园样板房二

88
ROYAL GARDEN CLUB
天宝物华会所

95
ROYAL GARDEN SHOW FLAT
天宝物华样板房

103
KING VISION ANOTHER CITY CLUB
君华又一城会所

115
KING VISION ANOTHER CITY SHOW FLAT A
君华又一城样板房一

123
KING VISION ANOTHER CITY SHOW FLAT B
君华又一城样板房二

129
KING VISION ANOTHER CITY SHOW FLAT C
君华又一城样板房三

133
KING VISION ANOTHER CITY SHOW FLAT D
君华又一城样板房四

140
XIYUE KINGKEY TOWN SHOW FLAT A
西粤·京基城样板房一

145
XIYUE KINGKEY TOWN SHOW FLAT B
西粤·京基城样板房二

151
GEMDALE LAUREL BAY SHOW FLAT
金地长青湾样板房

159
RSUN BOUTIQUE HOTEL
弘阳精品酒店

168
RSUN THE LAKEFORT BOUTIQUE HOTEL
弘阳三万顷精品酒店

172
YI TIAN GROUP OFFICE BUILDING
益田集团写字楼

180
SUNING OFFICE BUILDING
苏宁写字楼

186
LUNENG OFFICE BUILDING
鲁能集团办公楼

189
INTERNATIONAL DESIGN ASSOCIATES
国际设计联盟机构

190
STARWINNER
星胜石业

192
UNIVERSAL MARBLE & GRANITE GROUP
环球石材

194
PUYUAN DECOR
璞源软饰

195
HUA HUI CHENG
华汇诚

196
ORIENTAL CHIC
坤立家具

LIFE IS DESIGNABLE

人生是可以设计的

Art comes from life. In my boyhood, I still remembered that I could do some sketch all day in a raft on the river, or under the shade of tree, to draw the beautiful motherland's land, with a pot of water and two buns. The old house and the inverted image from water made a perfect harmony in my painting, suddenly a group of white birds flew to the frame, and my brush caught this moment. And this painting was praised by the fishermen behind. Since then, I had my own direction of life; any business practices can't shake my love and determination for art.

Unexpectedly, the social appeal of design art is so powerful, once the person attaches importance to the design and sets up his own brand firstly, both social and commercial benefits are expected to gain. Design Art also gives me joy and happiness for my life, family and career. There are three stages for my design career:
Family artistic influence in Childhood and the formal studying in Guangzhou Academy of Fine Arts;
A designer of Hong Kong De Sheng Design & Engineering Company at the beginning period of the Chinese reform and opening up;
A founder of design company, and establish DHA.

艺术源于生活。在我的少年时代，祖国的大好山河是那么美丽动人，一壶水两个馒头，在树荫下，在河边的竹排上，我就能写生一整天。画中老房子与水中的倒影相映成辉，突然一群白鸟飞来，这个瞬间被我牢牢抓住，通过画笔留存下来。专注之际被身后渔民的一阵笑声惊醒，"画得好像喔，哈哈……"从此，我为自己设计了人生的方向，什么样的商业行为都没有打动我喜爱艺术的决心。

没想到的是，设计艺术竟有这么强大的社会感染力，谁重视了设计，谁就最先建立了自己的品牌，社会效益与经济效益就有望双赢。设计艺术也给我的人生、家庭、事业带来了快乐和幸福。我的设计生涯，经过了以下三个阶段：
幼年家庭艺术氛围熏陶到广州美院正规学习；
中国改革开放初期任香港得盛设计工程公司设计师；
创办自己的设计公司及组建DHA国际联盟设计机构。

RICH'S CITY CLUB

君华新城会所

中山

As a designer, when we cooperate with the property developer, we not only should have a good understanding of the space, also have another task: helping customers to find the right theme story and make it the master of the space, allowing customers to experience comfortable and quiet feeling.

作为设计师，在与地产开发商合作时除了对空间的理解，还有一个任务，就是帮客户找到适合的主题故事，找到一位主角，让它成为室内空间的主人，让我们的客户在体验的过程中感到舒适和宁静。

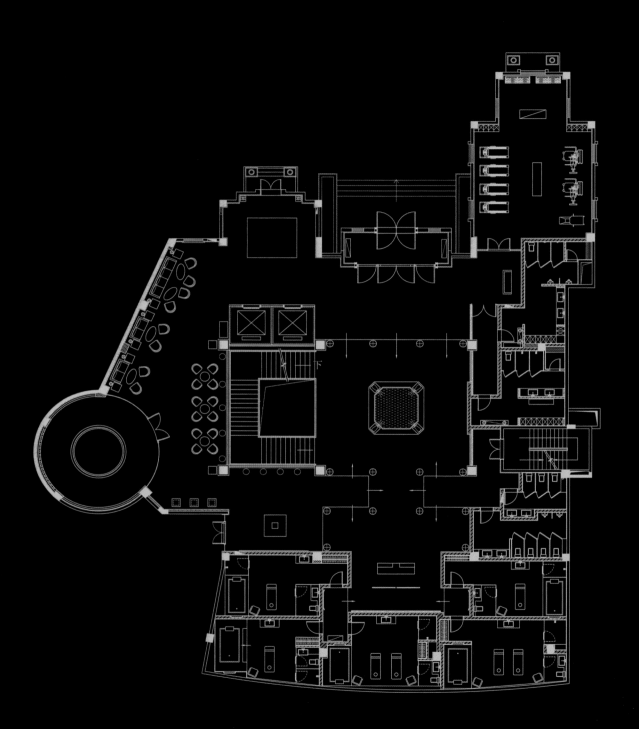

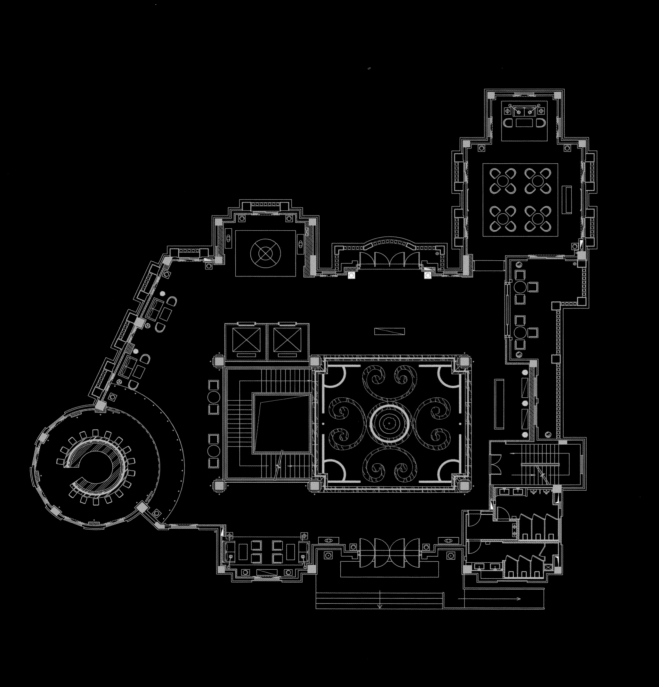

I have had a skilled artistry, but when I stop to study the means of artistic expression, it seems that I am a green hand.

Henri Matisse

我早已达到技精艺熟,可是如今我再研究自己的表现手法时,似乎觉得自己刚开始学习。

——马蒂斯

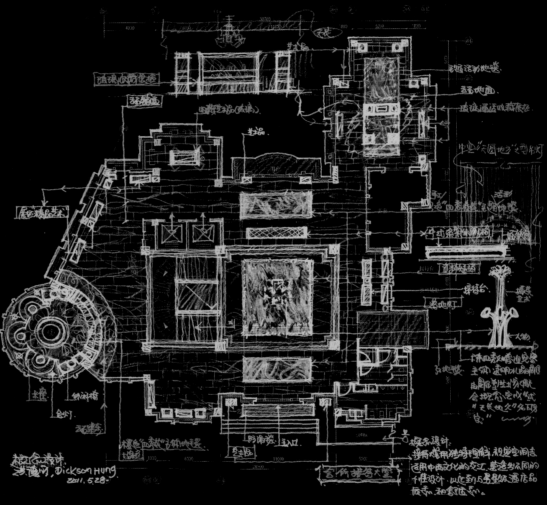

Seeking pleasure from works.
PaulCézanne

应从习作中寻求快乐。
——塞尚

Attempt to experience, refine the design concept of diversification is to keep the key element of the innovative design, which is creative source of meeting the different requirements of different customers.

大胆尝试、亲身体验、提炼多元化设计元素是保持概念创新的基石,是满足不同客户不同要求的创意来源。

KINGKEY TIMEMARK SALES CENTER

京基滨河时代售楼中心   深圳

Life is not lack of beauty, but lack of the eyes to find beauty.

Auguste Rodin

With economic development of the beginning of Chinese reform and opening up, the real estate of mainland followed Hong Kong's marketing model to do demonstration units for sale, which meant the start of interior space design industry. With the 20-year design experience in the design real estate projects, we still insist in this faith: the house price and demand of beauty are changing, but our innovative concept will never be changed.

生活中不是缺少美,而是缺少发现美的眼睛。

——罗丹

从中国改革开放发展经济开始,从内地学习香港房地产做示范单位销售开始,我们的室内空间设计也随之开始了。30年地产楼盘设计经验,我们不变的信仰是:楼价会变,人们对美的需求会变,但我们创新的理念一直没变。

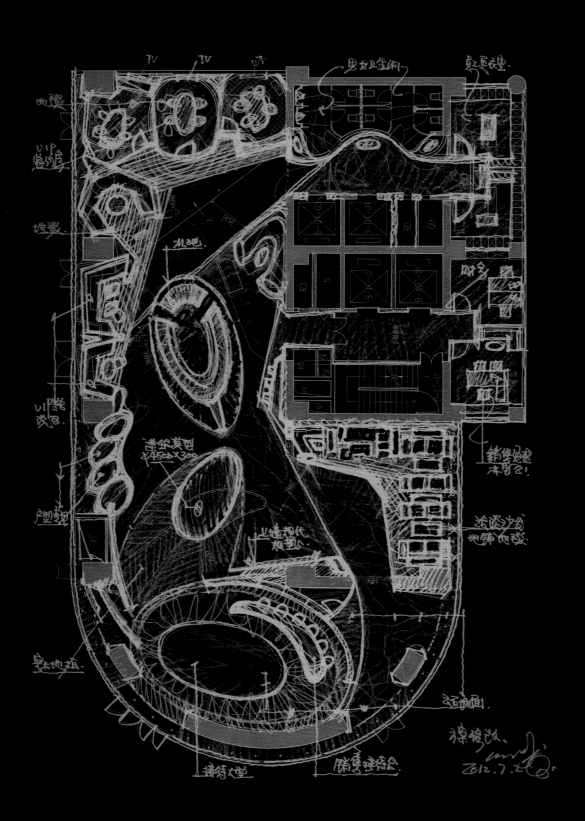

VIP Householder Lift Lobby. The designer wants to create a luxurious and comfortable environment, using natural Italian marble in the reception walls, which is very exquisite and moving as a Chinese ink and color painting.

贵宾住户电梯大堂。营造豪华舒适的环境,接待背景采用的意大利天然大理石,像是中国彩墨画般精美动人。

KINGKEY TIMEMARK
SHOW FLAT A

京基滨河时代样板房一 深圳

Perfection is not a small detail, but pay attention to detail can be perfect.

Michelangelo

For Design, detail plays an important role to be perfect, which can bring us spiritual pleasure.

完美不是一个小细节,但注重细节可以成就完美。

——米开朗基罗

设计强调用细节得到完美,它能给我们带来精神上的愉悦。

KINGKEY TIMEMARK SHOW FLAT B

京基滨河时代样板房二

30 years ago, at the beginning of my entering the design industry, I communicated with the builders, and this experience is useful for me to understand the workmanship, material characteristics and other basic elements. Nowadays, the application of materials and technology in the space design, allows us to play good role in the beauty of imagination.

30年前,入行之初我在工地现场与施工师傅们的沟通,让我对精细工艺、材料特点等基本要素有了更深刻的理解。今天,材料、科技在设计空间中的运用,能够让你发挥想象空间之美。

CINEASTE FOUR SEASON GARDEN CLUB

益田影人四季花园会所　　北京

Art is far less important than life, but life is boring without art.

　　　　　　　　　　　　　　　　　　　　　Ernst Mach

The project is located in Huairou, the largest film base in Asia. We take theme film as design concept, telling the lives of film makers and artists.

艺术远没有生活重要，但是没有艺术，生活是多么乏味呀！

　　　　　　　　　　　　　　　　　　　　——马赫为

位于北京怀柔亚洲最大的影视基地，益田集团影人四季花园会所以电影主题为概念，讲述电影人的生活。

CINEASTE FOUR SEASON GARDEN SHOW FLAT A

影人四季花园样板房一

Life is creativity, what you would like it to be, what it can be.

Tina Seelig

This is a movie star's home; we create a dream space, with female romantic and delicate emotions.

生活就是创意,你希望它是什么样,它就能是什么样。

——蒂娜·齐莉格

电影表演明星的家,女性浪漫而细腻丰富的梦想空间。

CINEASTE FOUR SEASON GARDEN SHOW FLAT B

影人四季花园样板房二

北京

Art is nature and man.

Francis Bacon

This is a director's house; we create this space to narrate a story of a film maker, who is dedicated his life to the film industry. And the living room is the communication space, sharing with friends.

艺术就是自然与人。

——培根

导演的家,讲述一位一生执着于电影事业创作的人,客厅成为可以与朋友共享的交流空间。

ROYAL GARDEN CLUB

天宝物华会所

扬州

Inspiration is exact "the reward of staunch labor".

Ilya Yafimovich Repin

Chinese always talk about Feng Shui philosophy, the customer requires us to reflect this relationship in this space, making full use of the five elements' relationship (metal, wood, water, fire and earth). Mutual generation of five phases: metal generates water, water generates wood, wood generates fire, fire generates earth, and earth generates metal. Mutual restriction of five phases: metal restricts wood; wood restricts earth, earth restricts water, water restricts fire, and fire restricts metal. We made succeed in this project, and the houses sold well, and we became friends.

灵感不过是"顽强的劳动而获得的奖赏"。

——列宾

中国人讲风水，我们这位客户希望在室内设计体现出这一关系，将"金木水火土"相生相克的元素考虑在空间之中。五行相生：金生水，水生木，木生火，火生土，土生金。五行相克：金克木，木克土，土克水，水克火，火克金。之后，客户的楼销售得很不错，我们也成了朋友。

ROYAL GARDEN SHOW FLAT

天宝物华样板房

 扬州

Eyes are the windows to the soul.

Da Vinci

Show Flat is a place to demonstrate and is the core element for sale. The designers should use their own experience to show different life space.

眼睛是心灵的窗户。

——达·芬奇

样板房是展示、销售楼盘的核心,设计师要用自己的经验去展现不同的生活体验空间。

KING VISION ANOTHER CITY CLUB

君华又一城会所

Color has the power like a bomb, and there must be a good way to control it.

Chang Dai-Chien

色彩像炸弹一样具有威力，可是也像炸药一样，要有很好的办法控制它。

——张大千

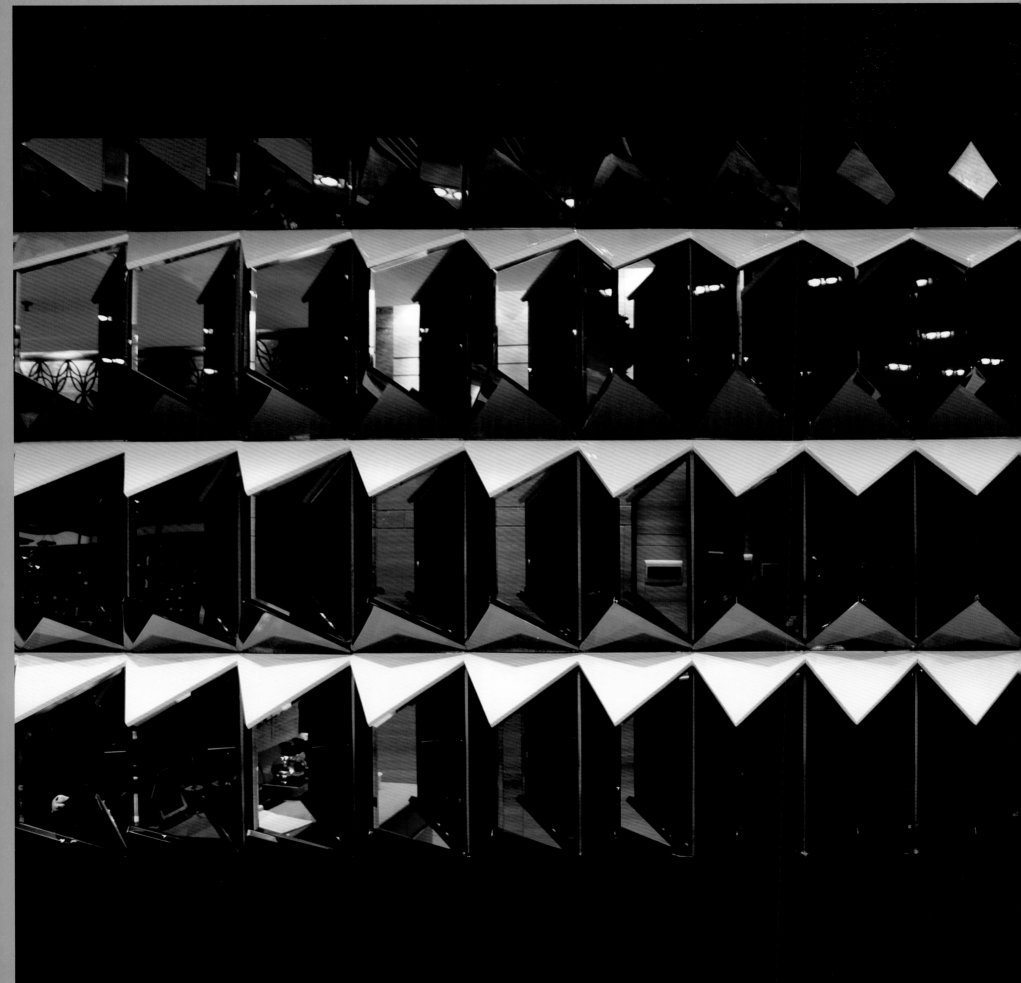

The design concept of the club is inspired by the "Queen Mary II", the largest and most luxurious yacht around the world, including the red and black chimney, white iron railings, natural wood floors, luxurious crystal chandeliers and powerful propeller. The blurred lights circle through the cabin windows onto the sea after dark. The light and shadow can provoke my rich spatial imagination.

会所灵感来自于世界最大最豪华的"玛丽皇后二号"邮轮，红色加黑色的烟囱，白色铁栏杆，天然实木地板，豪华水晶吊灯，还有那动力吓人的螺旋桨等。入夜之后迷离的灯光总会透过船舱的圆形窗户投射到海面上，这流光溢彩的倒影激起我丰富的空间想象力。

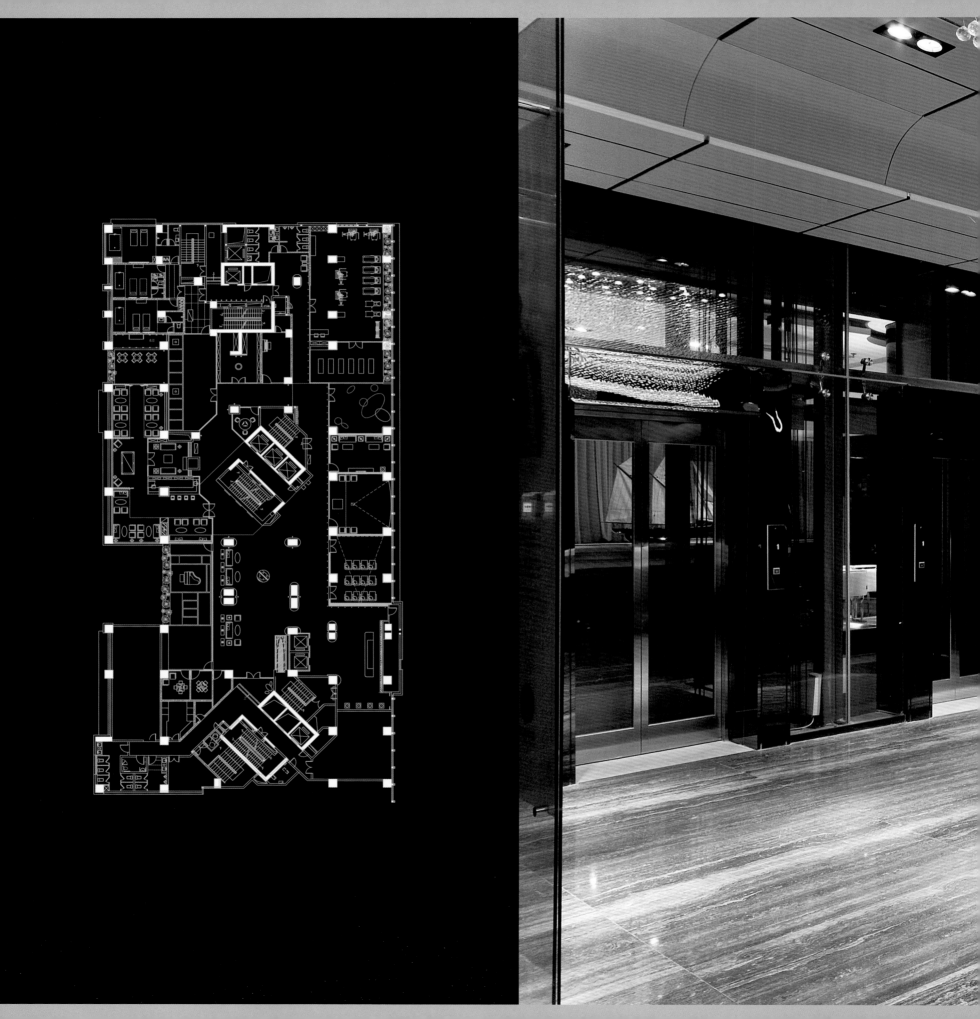

KING VISION ANOTHER CITY SHOW FLAT A

君华又一城样板房一

Creativity is a child; we need take an objective assessment to our own ideas, we all think them are the best.

Tina Seelig

Failure is not terrible, try once again; we always make a careful assay of project after accepting every commission. This space can be designed as a two bedroom and one living room structure, or a penthouse, or a villa. What designers need to do is identifying the spatial characteristics, and coordinate the relationship between the details and the whole space.

创意就像孩子,因为每个人都觉得自己的想法是最好的,所以我们要客观地评价自己的想法。
——蒂娜·齐莉格

尝试再尝试,失败并不可怕,我们接受每一位客户的委托设计后,都会认真分析,这个空间可能是二房一厅,或一个复式,也可能是一栋别墅,作为设计师需要找出空间特性,协调好细节与大格局的关系。

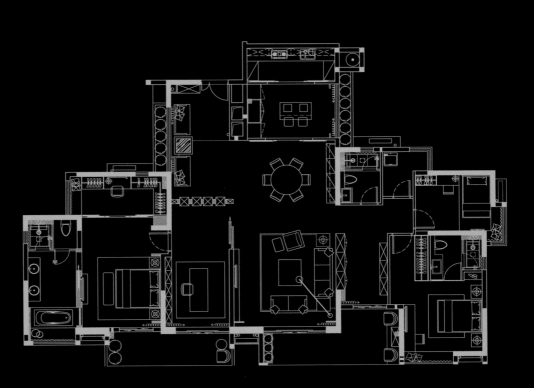

KING VISION ANOTHER CITY SHOW FLAT B

君华又一城样板房二

 广州

Creation is the return to nature.

Antonio Gaudi i Cornet

For designer, maintaining the natural feelings of life, making a moderate exaggeration in art to get a harmonious fusion is very important. This is key analysis element that designers do for real estate developers in terms of sales.

创作就是回归自然。

——安东尼奥·高迪

怎样保持设计师对生活的自然感受,同时在艺术上适度夸张,点到要处,这是我们为地产开发商在销售方面分析的要素之一。

KING VISION ANOTHER CITY SHOW FLAT C
君华又一城样板房三

 广州

Creativity is hiccups, it is difficult to stop once it starts, seize the inspiration and the opportunity.

Tina Seelig

创意就像打嗝，因为它一旦开始就很难停下来，所以要抓住灵感，不要错过了时机。

——蒂娜·齐莉格

In the hotel apartment, the show flat project, taking the one bedroom one living room and the Single-family villa for example, the skeleton of the space is the building itself; the designer is a tailor, tailoring the space from whole to parts, and combining and repeatedly studying, eventually creates an elegant and comfortable living environment.

酒店公寓、样板房,从一房一厅到独栋别墅,建筑是空间的骨架,设计师则像一个服装裁剪师一样,将空间从整体到局部分解、组合,反复推敲,最终形成优雅舒适的居住环境。

KING VISION ANOTHER CITY SHOW FLAT D

君华又一城样板房四

Any irresponsible inspiration doesn't mean long-term effort.

Auguste Rodin

Grasp the small space means the designer has the ability to manage the relationship of whole space. We often think about the relationship between the point and the surface, between the line and point of the space to make the design have much more vitality.

任何倏忽的灵感事实上都不能代替长期的功夫。

——罗丹

把握住细小的空间就能把握住大空间的相互关系，我们经常在点与面、线与点的空间滑道上来回思索，让设计更有生命力。

XIYUE KINGKEY TOWN SHOW FLAT A

西粤·京基城样板房一 　湛江

Many artists just accept only one painting method and accuse other methods, which lead to their failure. The artist should study all of the painting methods in the right way to maintain their own uniqueness, but not follow one artist.

Eugène Delacroix

许多艺术家的失败，仅仅是因为他们只接受一种画法，而指责其他所有的画法。必须研究一切画法，而且要不偏不倚地研究；只有这样才能保持自己的独特性，因为你将不会跟着某一个艺术家跑。

——德拉克罗瓦

Each real estate developer has a different selling position, so they have different design requirements. I am very grateful to such challenges for creation. Designers should accumulate continually innovation idea in the actual work, and then realize the dream of success.

每个房地产开发商都有不同的销售定位和不同的设计要求，虽说给我们设计师带来了不一样的挑战压力，但我非常感谢我们的客户给了我们这么多的创意机会，设计师只有在实际工作中不断积累创新，才能实现成功之梦。

XIYUE KINGKEY TOWN SHOW FLAT B

西粤·京基城样板房二

湛江

Artists create a painting with the brain rather than a hand.
　　　　　　　　　　　　　　　　　　　　Michelangelo

Focus on one thing in a lifetime, in my heart the exact thing is design art.

艺术家用脑，而不是用手去画。
　　　　　　　　　　　　——米开朗基罗

一辈子专注一件事，那就是我心中最爱的设计艺术。

GEMDALE LAUREL BAY SHOW FLAT

金地长青湾样板房   沈阳

When customers praise our design works, I feel grateful to their recognition and trust. The customer's understanding and support is key element to make sure the design completeness in the process of realizing the conceptual design.

当客户赞扬我们的设计效果不错时，我的第一个反应是感谢客户的认可和信任。在实现概念设计的过程中，客户的理解和支持才是设计不走调的关键。

RSUN BOUTIQUE HOTEL

弘阳精品酒店

 南京

The gifted person always work most in mind when they work least ostensibly, because they are making a preliminary sketch to mature, and then these ideas will be subsequently expressed.

Da Vinci

有天资的人,当他们工作得最少的时候,实际上是他们工作得最多的时候。因为他们是在构思,并把想法酝酿成熟,这些想法随后就通过他们的手表达出来。

——达·芬奇

Hongyang boutique hotel is very simple and clean; the designer wants to show its personality in design specifications, using local traditional crafts "Wenjin" pattern, handed down from two thousand years ago. The artistic personality of the hotel is expressed by the Chinese and Western design techniques.

南京弘阳精品酒店,简约干净,在规范设计中寻找个性。设计中运用了当地 2000 多年前的传统手工艺"文锦"图案,中西合璧的设计手法增添了酒店的艺术个性。

RSUN THE LAKEFORT BOUTIQUE HOTEL

弘阳三万顷精品酒店 无锡

The neo-classical boutique hotel was under construction, at that time, the Italian designers and American designers of DHA International Design Alliance team not only made us to have a deep understanding of European culture, but also gave us a lot of valuable advices. Centered around the key theme, we used the "classical elements, modern style" to tell the interesting story of each space.

正在施工的新古典精品酒店，DHA 国际设计联盟团队的意大利设计师和美国设计师让我们对欧陆文化有了深刻的理解，同时也给了我们很多宝贵的意见。围绕着主题，我们用"古典元素、现代手法"讲述每个空间的有趣故事。

The painter, with surreal eyesight, creates works to arouse people's imagination.

Pablo Picasso

画家的眼睛,可以看到高于现实的东西,他的作品就是唤起人们的想象。

——毕加索

YI TIAN GROUP OFFICE BUILDING
益田集团写字楼

Creativity is a mirror, which only reflects its surrounding environment, so we should absorb some freshest ideas to change environment constantly,

Tina Seelig

创意就像一面镜子，因为它只能反射自己周围的环境，所以要不断改变环境，多吸收一些新鲜的理念。

——蒂娜·齐莉格

In my opinion, the interior space is used by people, and the office space is the most important space for white-collar workers working and living every day, and the theme of the space should not be too serious and depressive. In this project, we break the traditional pattern to create a relaxed space; the employees can feel as comfortable as at home. Such office environment greatly enhances the work efficiency, and will be a good measure to keep talent.

室内空间是给人使用的，写字楼是白领们一天最主要的工作生活空间，不应过于严肃而显得压抑。打破传统格局，让员工感到放松自在，像在家一样地舒服，这样的办公环境会大大提升工作效率，当然也能留得住人才。

SUNING OFFICE BUILDING

苏宁写字楼

南京

People always think that artistic creation for me is very easy, this is wrong, because no one can spend so much time and effort on the composer like me.

Wolfgang Amadeus Mozart

The Suning Group office building is located on the top floor of the five-star Sofitel Hotel in Nanjing. The project makes us realize the extended significance of design, using design language to show corporate brand image and the development momentum of Suning Group.

人们认为：我的艺术创作轻而易举。这是错误的。没有人像我那样在作曲上花费了如此大量的时间和心血。

——莫扎特

苏宁集团写字楼位于南京市五星级酒店索菲特酒店的顶层。这一作品让我们深感设计之外的成功，用设计的语言表现出企业品牌形象，展现出企业的发展动力。

LUNENG OFFICE BUILDING

鲁能集团办公楼

济南

Imagination is much more important than knowledge, knowledge is limited, while imagination summarizes everything in the world, promotes the progress. Imagination is the source of knowledge.

Albert Einstein

This is a 16-storey corporate office building project; we use simple and clear space language to define the relationship of clean space and complex structure.

想象力比知识更重要，因为知识是有限的而想象力概括着世界上的一切，推动着进步，是知识的源泉。

——爱因斯坦

一座 16 层的企业写字楼设计，我们用简单明确的空间语言界定了干净与复杂的关系。

INTERNATIONAL DESIGN ASSOCIATES

DHA
INTERNATIONAL DESIGN ALLIANCE
國際設計聯盟機構

酒店总体策划顾问
The hotel overall planning consultant

酒店总体室内设计
The hotel interior design

酒店环境灯光设计
The hotel lighting design

酒店总体建筑规划
The hotel overall planning

酒店艺术设计配饰
The hotel decoration art

酒店家具配套设计
The hotel furniture design

A: 深圳市福田区益田路 3013 号南方国际广场 B 座 1018 室
Room 1018, Southern International Plaza B, No.3013, Yitian Road,
Shenzhen, China   T: 0755-82823028   E: dha_sz@163.com

A: 上海市长宁区延安西路 777 号 1501 室
Room 1501, No.777, Yan'anxi Road, Changning District,
Shanghai, China T: 021-52388130 E: dha_sh@163.com

STARWINNER

星胜石业

STARWINNER
星胜石业

深圳市星胜石业有限公司 — 营销中心
A: 深圳市南山区华侨城创意园 A4 栋 502B
T: +86 755 86096198   F: +86 755 86095168
E: starwinner@china.com

SHENZHEN STARWINNER STONE CO.,LTD
A: 502B ,A4 BLOCK ,LOFT,OCT,SHENZHEN,CHINA
T: +86 755 86096198   F: +86 755 86095168
E: starwinner@china.com

从业 20 年，全球 12 座矿山中国总代理

UNIVERSAL MARBLE & GRANITE GROUP

环球石材

UMGG | LOFT
UNIVERSAL MARBLE & GRANITE GROUP

地址：广东省东莞市长安镇厦岗管理区
电话：400-680-1987
Address: Xiagang, Precinct, Chang'an Town,
Dongguan, Guangdong Province, China
Tel: 400-680-1987
www.umgg.biz   www.stonexp.com

PUYUAN DECOR

璞源软饰

深圳市璞源软饰材料有限公司
地址：深圳市福田保税区福年广场 B3 座 625
电话：0755-23914363
传真：0755-23914363-802

Address: Room 625, Block B3, Funian
Square, Futian Bonded Area, Shenzhen
Office: 0755-23914363
Fax: 0755-23914363-802
E-mail: pydecor@163.com
Http://pydecor.com

HUA HUI CHENG

华汇诚

华汇诚马赛克
MOSAICO ART

深圳市华汇诚建材有限公司
地址：广东省深圳市福田区福强路乐安居新洲家居装饰
材料市场一楼 108 号
电话：0755-83454227 25685557
传真：0755-82046609

SHENZHEN HUA HUI CHENG DECORATION CO.,LTD
Address:No.108,1/F,Xinzhou Home-Design-Material Market
Roman Joy,Fuqiang Road,Futian District,Shenzhen,Guangdong
Tel: 0755-83454227   25685557
Fax：0755-82046609
E-mail：cycmosaic@163.com
Http：//www.szmosaic.com

ORIENTAL CHIC

坤立家具

MURRO
by Oriental Chic

东莞坤立家具有限公司
地址：东莞市厚街镇白濠村北环路第三工业区
电话：0086-769-85939386
传真：0086-769-85939396

Address: No.3 Industry Area, Beihuan
Road, Baihao Estate,
Houjie Town, Dongguan
Tel: 0086-769-85939386
Fax: 0086-769-85939396
E-mail: innocasa@vip.163.com
www.orientalchicsofa.cn

To My Parents 谨以此书献给我的父母亲

洪德成 生活 艺术

LIFE & ART

DICK
SON
HUNG

LIFE & ART
生活 · 艺术

I have a dream, a passion for design art, and as if I was born for it. "Design Art" — Why do I have such an understanding and give it such a name? In China, space design is a commercial activity in contemporary, and the influence of customer's subjective opinions on a project design can account for 70%. While the designer have the right to change the construction team as its quality and management problem In Europe. It's totally different from China. During my 30-year design career, there is seldom a customer, especially real estate developer, who can accept completely my creative idea, which makes me feel deeply distressed for the big design fee.

Compared with other hobbies, painting and photography are my true authentic art design language, so my paintings can be seen everywhere in my interior space design projects. Sometimes I use the modern abstract original oil painting, marble mosaic, the original natural wood and carved crystal glass and other techniques to decorate the space and make the story of space vividly.

I like photography; I take photos for most of the interior design works. I travel to India, Nepal, Yunnan, Sichuan-Tibet and other places to photo with photography enthusiasts in Holidays; we drove off-road vehicles on the winding mountain road at an altitude of 5000 meters, made telepathic communication with the villager with a camera in our hands, and experienced the local lives and culture. Look the world from another angle, experience another way of life, this is my favorite way to relax.

I also like a small weekend party with friends, with the same love for art; some of them are professional painters or university professors. We describe the creative process, discuss the painting techniques and share the experience of expressing emotion by brush. Sometimes, I will co-author an original painting work with my first teacher — De Zhong, my elder brother.

我心中一直有一个梦，那就是对设计艺术的喜爱，好像我天生就有一颗对此而执着的心。"设计艺术"——我为什么这样理解和称呼呢？因为空间设计本身是商业行为，在当下的中国，客户的主观意见对一个设计项目的影响力可以占到 70%。而在欧洲，设计师可以因施工队素质及管理问题，向投资方提出替换该队，但在中国行不通。我从业 30 年，还没有遇到过一个完全任我发挥创意的客户，特别是地产商的项目，走样的设计常常让我为客户付出的高额设计费深感不值和痛心。

相比之下，绘画、摄影则是我真正原汁原味的设计艺术语言。基于这个原因，在我的室内空间设计中，我的绘画也无处不在。有时用现代抽象油画的原作；有时用大理石马赛克表现；或用天然原木和水晶玻璃雕琢等手法来点缀……它们会让空间故事更加生动。

我还喜欢摄影，公司的室内设计作品大部分是我自己拍摄的。假期与摄影发烧友相约去印度、尼泊尔、云南、川藏线等地旅游，驾着越野车在海拔 5000 米的盘山公路上，用手中相机与纯朴的村民进行心灵沟通，体验当地的生活文化。换个角度看世界，换个方式生活，这样的放松方式是我的最爱。

我也喜欢和发小的周末聚会，同样热爱艺术的他们现在已是专业画家、大学教授，在说笑畅谈中，我们讲述各自的创作过程，研讨绘画技法，交流用画笔表达情感的心得。有时，我还会和我的启蒙老师——大哥德忠合作，共同创作一幅油画原创作品。

My parents name my brother and me Dezhong and Decheng require us to conduct ourselves and do things loyally.

父母为大哥和我起名为德忠、德成，希望我们做人做事记住忠诚。

CONTENTS
目录

6~47
绘画

48~181
摄影

182~189
交流

190~193
朋友眼中的洪德成

194~195
母爱，在艺术的起点

RICH IMAGINATION

丰富想象

I intended to be a painter, while the fate made me indissoluble bond with space art. Fortunately, there is no boundary in art. Painting can cultivate your aesthetic and manual skills. A spatial concept formed in the brain, and then the sketching can fully demonstrate its details, including the structure, proportion, color, light and shadow, texture, process and so on. In my opinion, the computer can only represent the concept of "form", but can not express the "god" of my heart.

原本想当个画家,命运却让我与空间艺术结下了不解之缘,好在艺术不分国界。绘画艺术能培养你的审美和动手能力。当一个空间概念在大脑中形成后,手绘能将其细节充分表现出来,包括结构、比例、色彩、光影、质感、工艺……在我看来,电脑只能代表概念的"形",但不能表现我心中的"神"。

No person who is not a great sculptor or painter can be an architect. If he is not a sculptor or painter, he can only be a builder.

John Ruskin

In Architecture and interior design major of the United States University, one has to learn the architecture courses at first two years, laying a good foundation, then started professional learning. The knowledge of architects and interior designers complement with each other, such knowledge is very important.

As a designer, I have been exploring how to combine painting art with visual effects, and how to apply it to my understanding of exploration.

建筑师必定是伟大的雕塑家和画家。如果他不是雕塑家和画家，他只能算个建造者。

——约翰·拉斯金

在美国，某大学的建筑学和室内设计学专业，前两年的课程都是学习建筑，打好基础之后，再分别进入更加细分的专业学习。建筑师和室内设计师的知识互补，是非常重要的。

作为设计师，我一直在探讨怎样将绘画艺术与空间视觉效果相结合，并且有效地运用到自己的理解探索中。

In recent years, I regained the brush again; friends said that it's hard for me to change the habit. The basic skill of painting comic book lets me take dictation directly on the paper about the impression of the life of the Shangri-La in Yunnan.

近年来我开始重拾画笔，朋友说这是我改不了的天性，从前画连环画的功底，今天仍能让我直接在宣纸上默写出对云南香格里拉的生活印象。

In order to convey the feelings of their experience to others, people should rekindle those feelings in mind, and express it through some external signs — this is the origin of the art.

Tolstoy

Nowdays, painting with brush is totally different from that of 30 years ago, the life experience recorded in mind, so I have the ability to release the inner language of true art with unscripted and direct expression.

一个人为了要把自己体验过的感情传达给别人，于是在自己心里重新唤起这种感情，并用某种外在的标志把它表达出来——这就是艺术的起源。

——托尔斯泰

今天拾笔画画和 30 年前大不一样，生活历程全记在脑里，不用草稿而直接用默写的能力来释放内心真正的艺术语言。

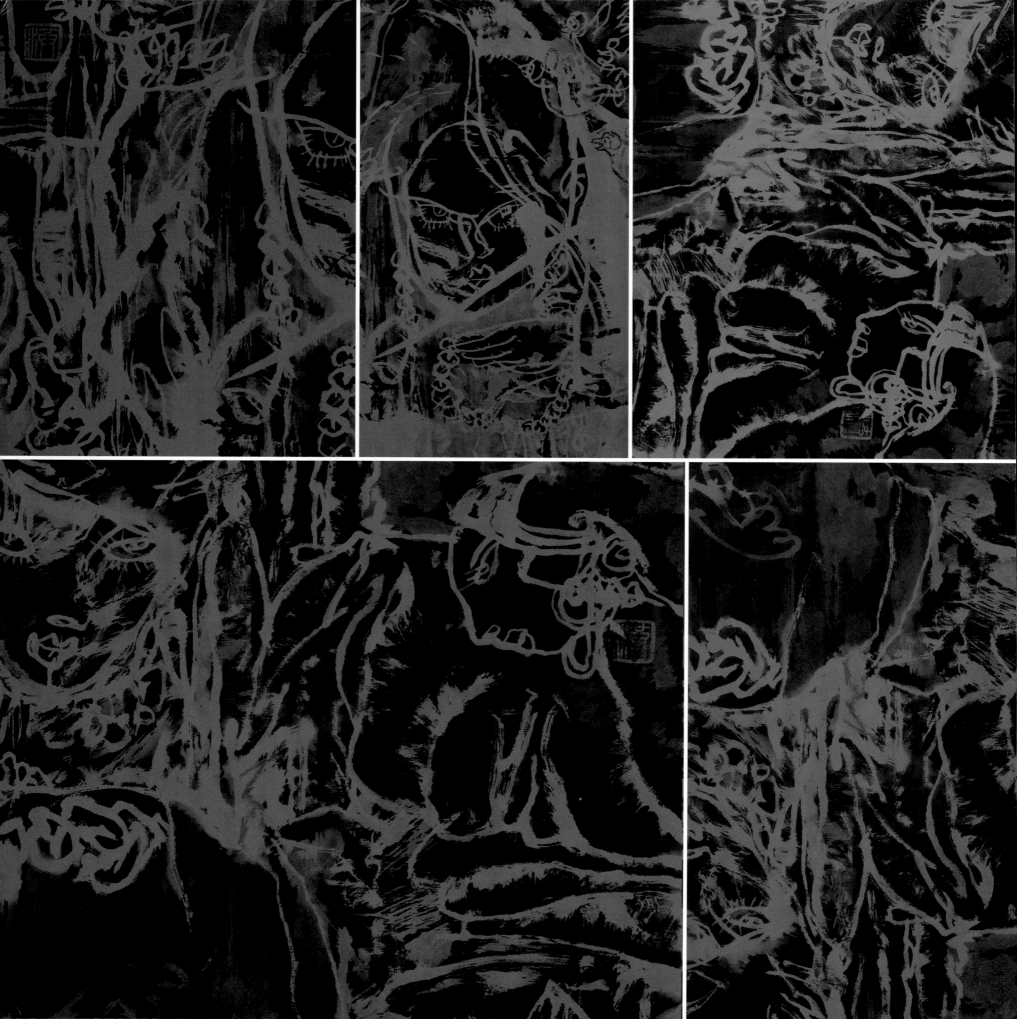

Dezhong, my elder brother and I not only have blood relationship, but we are like-minded friends of painting, which is most rare. What gratifies parents is we have sense of artistic, and we can have a bright future. Now, his abstract painting has formed his own unique characteristics. Sometimes, we discuss European abstract works together; we also co-author works, and some joint cooperation paintings can be seen in my interior design works.

我和德忠大哥既是亲兄弟，又是志趣相投的画友，这最为难得。父母在世时最欣慰的就是我们兄弟二人都有艺术细胞，未来有发展。现在，大哥的抽象油画已经形成自己的特色，别具一格。有时，我会和大哥一起探讨研究从欧洲带回来的抽象画派作品；有时，我们还会共同创作，在我的室内设计作品里，经常能看到我们联手合作的绘画作品。

SIZE: 200cm x 160cm

These two works remind me of the youth story 35 years ago. These two works are my painting assignments in Guangzhou Academy of Fine Arts. One is still life painting; one is Hainan Miao Village painting.

35年前的青春故事恍若眼前。这两幅作品是我学生时代在广州美院的写生习作，一张是静物写生，一张是海南岛苗族村落写生。

SIZE: 120cm x 58cm

SIZE: 150cm x 58cm

SIZE: 100cm x 100cm

The project is Zhongshan Junhua "World Wine Museum" theme club in Guangdong province, VIP room was designed with nostalgic movie concept. From the sketch, we can see characters and themes including the sculpture of Oscar Academy Award, Monroe, Gong Li, Avatar, Confucius, Gladiator, Waterloo Bridge. This 5mx4.5m piece of themed painting was finally produced by the 10mmx10mm marble mosaic.

广东中山君华"世界红酒博物馆"主题会所，VIP 怀旧电影概念设计。从草图中可以看到奥斯卡金奖雕塑、梦露、巩俐、阿凡达、孔子、角斗士、魂断蓝桥等人物和主题。这幅 5mx4.5m 的主题画最后用 10mmx10mm 的大理石马赛克制作完成。

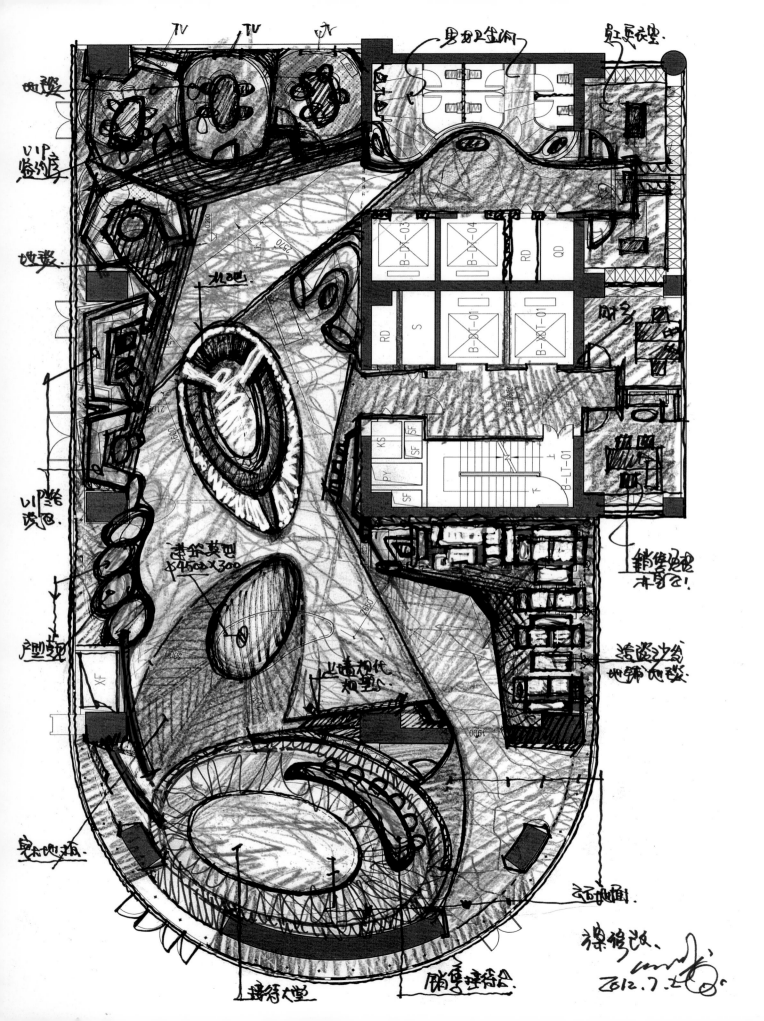

Compared to the computer, I am still used to using hand-drawing drafts to express my initial concept formation. The computer brings us great convenience, but hand-drawing can express your initial idea quickly and freely.

相比于电脑，我还是习惯用手绘草稿来表达我最初概念的形成。电脑能带来极大的方便，但手绘能更快速地随意表达你最初的构想。

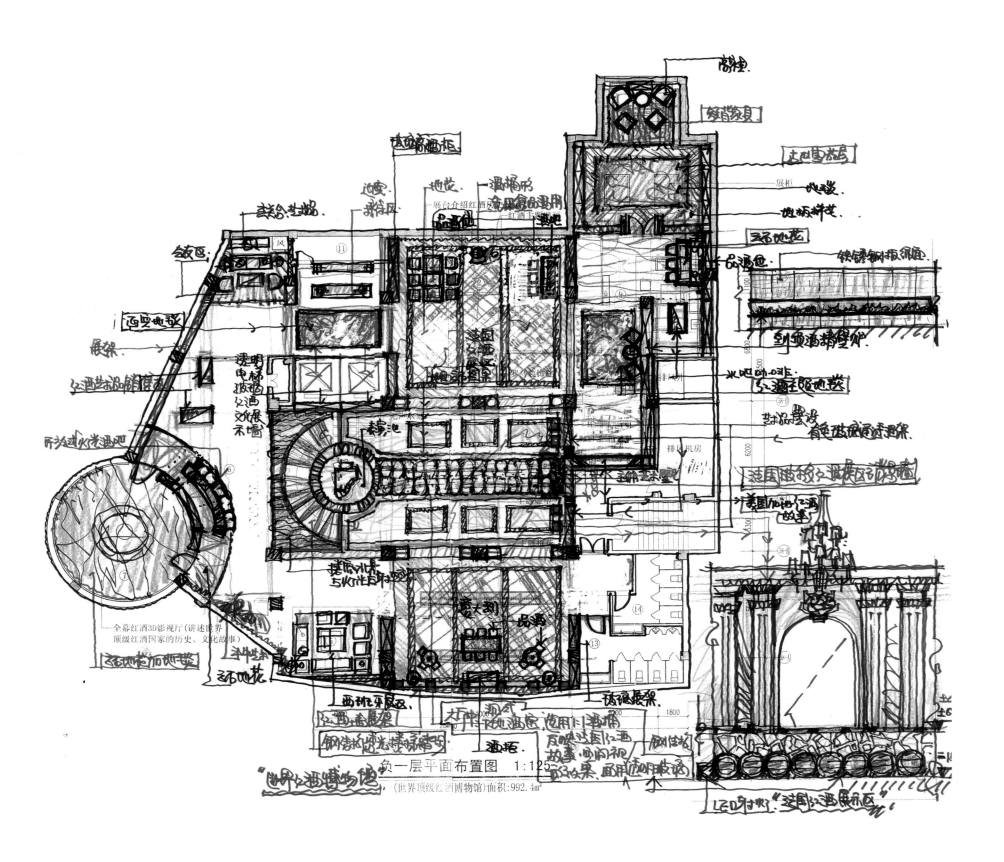

The inspiration of Guangzhou Luxury Yacht Theme Club is from the story of world's largest and most expensive yacht "Queen Mary Ⅱ". After carefully watching the documentary of the construction of a world-class yacht, I expressed heartfelt admiration to the investors, designers, construction workers and participants by sketching. Finally manuscript was engraved in the glass, to be one part of the club design. The projection and shadow of the night light on the sea gave me the inspiration for creating handmade carpets. The shape of crystal light is derived from the dolphin and yacht propeller.

广州君华豪华邮轮主题会所的灵感，来自于世界最大最昂贵的远洋邮轮"玛丽皇后2号"的故事。在认真观看了建造这艘世界级邮轮的纪录片之后，我通过手绘草图表达了对邮轮投资者、设计者、建造工人及参与者的衷心敬佩。手稿最后被刻在玻璃上，成为会所空间的一部分。夜间邮轮投射在海面的光影，给了我创作手工地毯的灵感。水晶灯的造型创意则来源于邮轮的螺旋桨和海豚。

Once the project needs using special materials and technology, I will go to the factory and work together with the workers to study how to achieve the best design effect. Excellent design project won't miss any detail.

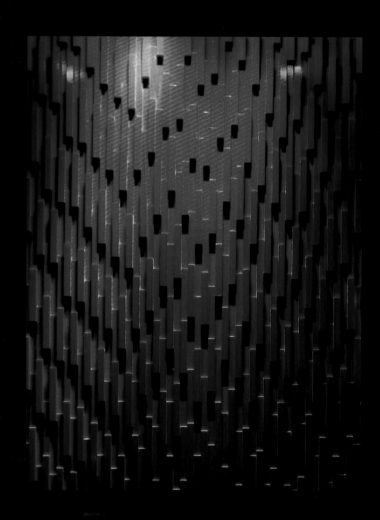

当设计中需要用到特制材料工艺时,我会去工厂和工人们一起研究怎样才能达到最佳的设计效果。精品,就是不放过任何细节。

CAPTURE THE INSPIRATION
抓拍灵感

I regard photography as musical notes, which gives me great relaxation to my mind, slowing it down to adjust the pressure of work. I capture the spirit, form and shadow of nature by the lens, to experience different lives of culture. And a lot of creativity is from the moment pressing the shutter.

摄影就像音符一样,给我的心灵带来了极大的放松,放平心态,调节我平时工作的压力。我会透过镜头去捕捉大自然中神、行、影的存在,体验不同的文化生活,很多创意都来源于按下快门的那一瞬间。

攝影環境

Durbar Square Kathmandu Nepal 尼泊尔加德满都杜巴广场

Nepal monk in burned corpse temple

尼泊尔烧尸庙里的苦行僧

The people in Nepal are very simple, with a strong sense of respect and protection for cultural relics; architectural wood and stone carvings are seen everywhere in people's real life.

在尼泊尔,当地人很朴实,人们对文物的尊重保护意识很强,建筑上的木雕石刻在人们的现实生活中随处可见。

This is a wood carving art village in Nepal, we can see the art sculptures with history of hundreds of years everywhere, what made me surprised is all art sculptures are preserved well.

尼泊尔的一个木雕艺术村，到处可见几百年前的艺术雕刻品，令我惊讶的是，它们都保存得如此之好。

An old man, bowed down with years, approached from far away, compared with a pair of majestic stone lions behind him, he was so strong. I captured this moment.

一位伛偻老者，远远走来，正巧经过一对雄伟的石狮，对比之下他是那么的坚强，我拍下了这一瞬间。

An ascetic man, sitting in the sun every day, stared at the blue sky. "Why?" We asked him. He said that although he had never been in an airplane, but there was a plane in his left arm.

一位苦行僧人,每天坐在太阳下凝望着蓝天。为什么?我们问他。他说,虽然我从来没乘过飞机,但是你们看,在我的左手臂上有一架飞机。

India village in Nepal, a villager is shepherding, I can feel her love and hope for life from her eyes, although we can't use language to communicate.

尼泊尔印度村,一位村民在牧羊,虽然无法用语言沟通,但从她的眼神中,我能感受到她对生活的热爱和希望。

In order to watch 5:00 a.m.'s sunrise, we should get up at 4:00 a.m. and finally I was courageous enough to reach the summit of the mountain with my hurt leg without accepting the advice of the companion. Unfortunately we missed the sunrise; I took this picture to memorize that experience.

凌晨5点要登顶看日出,于是4点起床,由于我的腿脚受过伤,同伴要我在山下等,最后我却凭着勇气登上了山顶。遗憾的是错过了日出,只此镜头,留作纪念。

In Nepal, such kind of house likes painting in my eyes, wood carvings in buildings, faded red bricks; roof tiles and yellow grass are the real authentic art of living material.

在尼泊尔,这样的住宅在我眼里就像一幅幅油画,那些建筑木雕,褪色的红砖,屋顶上的瓦片中还长着黄油油的草,真是原汁原味的生活艺术素材。

Such a sincere brotherhood.　兄弟情是那么的真诚。

In order to capture the millet on the head of the peasant woman, my camera should highly shoot from low angle, thanks for her cooperation.

为了抓拍这位农妇头顶谷子的镜头,我的相机必须高度向上仰拍,好在她很配合。

The eyes of the girl are so moved and pure. 那双眼睛是那么的动人、纯真。

In Nepal market, a beautiful local girl is selling dried fish. When she saw me raising my camera, she was shy, after I praising her beauty with gestures, I finally took this photograph in free demeanor, a painting feeling.

尼泊尔的菜场,一位漂亮的本地姑娘正在卖干鱼。看到我举起相机时她有些腼腆,我用手势夸赞她靓丽,终于让我拍下了这张神态自如的照片。油画的感觉。

I used the Nikon D3S digital camera with 70 to 200 lens to record this shepherd girl in early morning.
我用尼康 D3S 数码相机 70~200 镜头记录了这位早出放羊的小女孩。

The cattle of Nepal are white; farming tools were only seen in my childhood in rural China.

尼泊尔的耕牛是白色的,种田的工具我只有小时候在中国农村见过。

Farmer 农家汉

Mother Love 母爱

In India, the man, blowing the bamboo flute over the Ganges River reminded me the memory of the student hood: Chinese flute solo, "pastoral song."

在印度,恒河边吹竹笛的人,让我回到了学生时代。中国笛子独奏——"牧民新歌"。

Girl's smile 少女的微笑

Stone Sculpture of the India's Love Temple is vivid: sophisticated technology, beautiful designs. 印度的爱神庙石雕造型生动,工艺精细,图案优美。

An India girl 印度少女

131

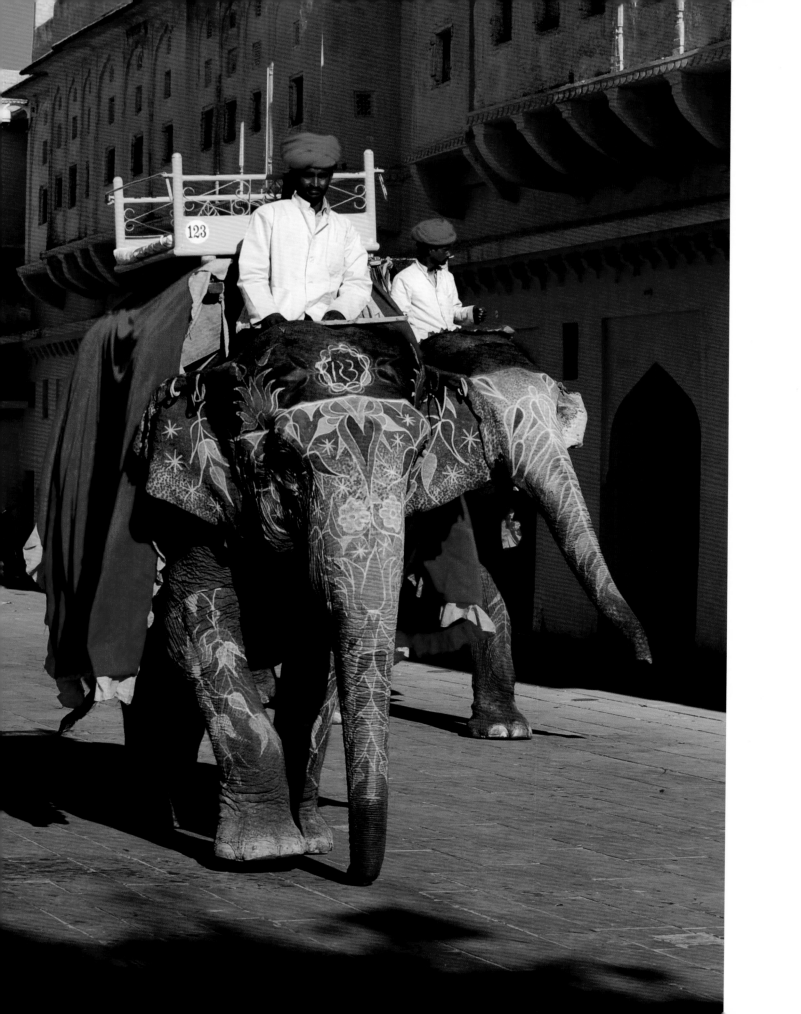

坐在石象上的少女
The girl, sitting on a stone elephant

143

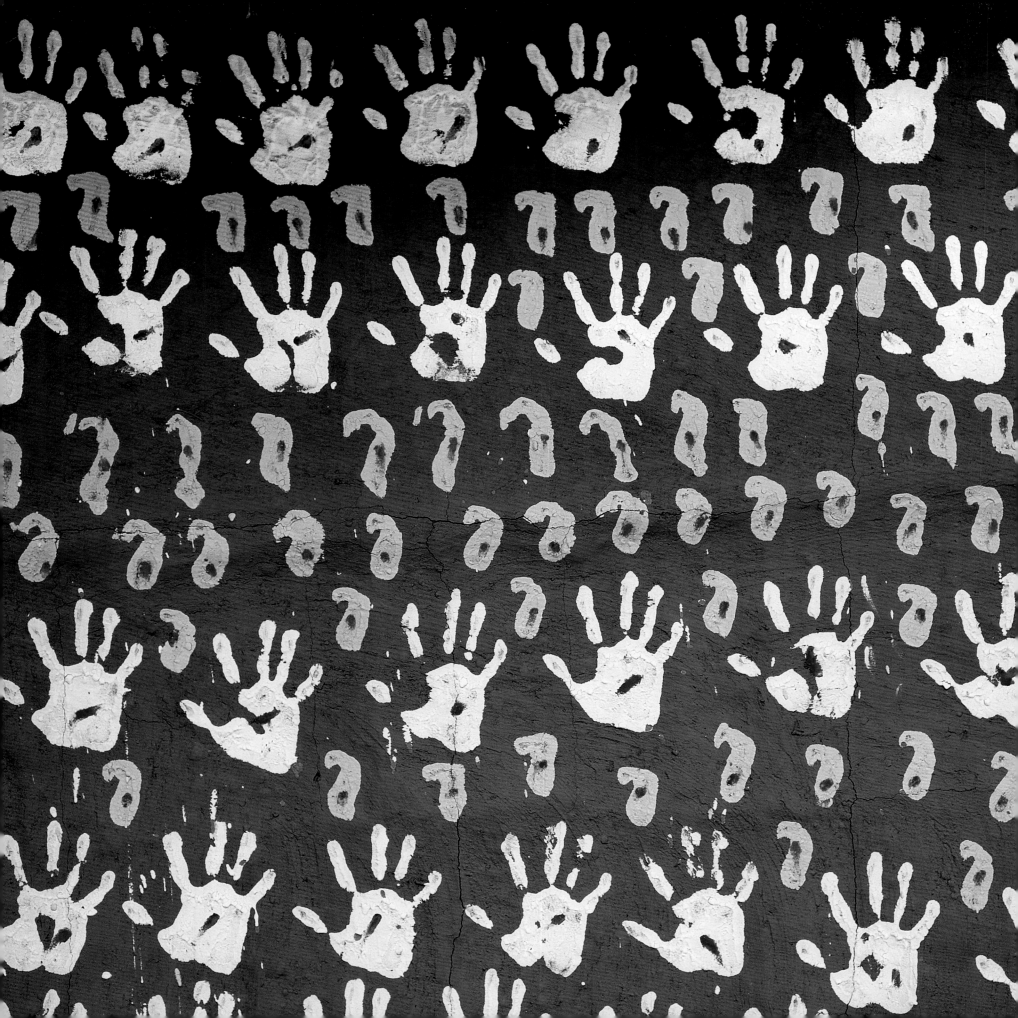

Sichuan KangDing Towen Tower
四川康定塔公

Watching the sunrise, overlooking the Tibetan cattle; the life, the nature's gift, is full of vigor under the sun.

山顶看日出，俯视藏民放牛，大自然赐予的生命在阳光的哺养下生机勃勃。

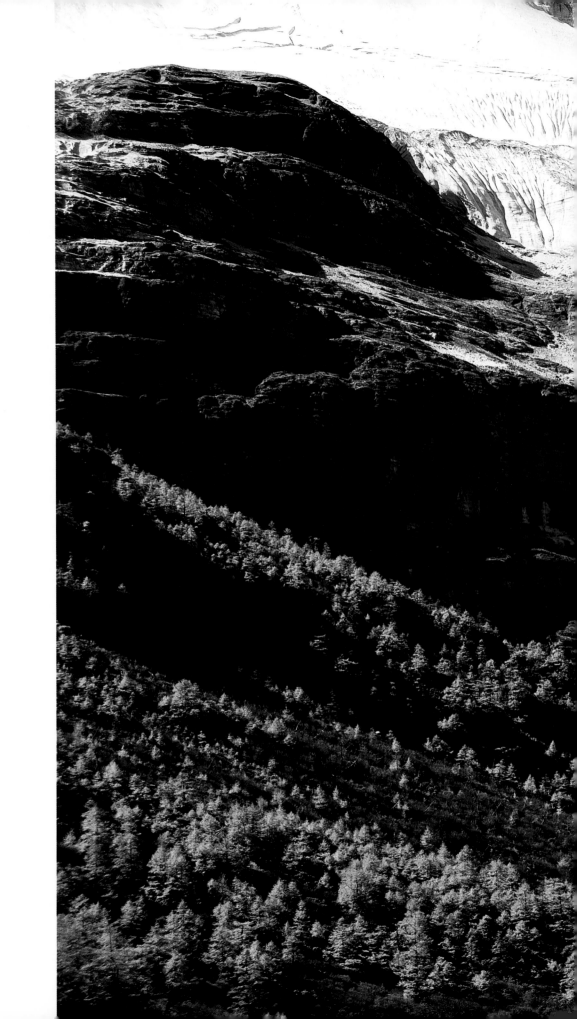

Yang Maiyong Snow Mountain is one of three God-mountains in Ya Ding, Sichuan Province. The landscape seems like a painting. Using 70 to 200 lens and double teleconverter lens to capture the snow-capped mountains, looking those magical details, ignites passion for art.

四川亚丁三大神山之一的央迈勇雪山,风景如油画一般。用70~200加上双倍增距镜头将雪山拉近,细看那些神奇的细节,心中燃起对艺术的热情。

The morning sun bursts the views of Sichuan Danba Tibetan village. 清晨日光用点射托起了四川藏族丹巴村落的美景。

In the mountain village, we sometimes went to the Tibetan villager house to have diner and drink buttered tea together.
在山里我们有时会去藏民家做客，吃饭，喝酥油茶。

It took three days and nights for the sea trip from Miami to the islands of the Bahamas.
从美国迈阿密经过三天三夜的海上航行,到达巴哈马岛国。

DESIGNED TO LOOK AT THE WORLD

交流看世界

"Communication" habit has been formed since I was a child. I like sharing with others, especially with my mother, no matter daily life or studying matters.

I still remember one sentence by my friend, Massimo Formenton, the Italian designer, he said: "Design has a soul." So I insist on communication studying to capture the soul of design art.

"交流"是我从小就养成的习惯,不管是生活上的还是学习上的事,我都喜欢与人分享,特别是我的母亲。

我的朋友意大利设计师玛斯莫·福尔蒙顿(Massimo Formenton)说过一句话让我至今难忘,他说:"设计是有灵魂的。"因为这句话,我一直坚持"交流"学习,捕捉设计艺术的灵魂所在。

IMPRESSION OF DICKSON HUNG
朋友眼中的洪德成

Shenzhen (China) International Trade Building is the landmark building of Chinese reform and opening up period, which is also a milestone for interior design industry development in China. Since Mr. Dickson Hung participated in the interior design of International Trade Building, his interior design career has been nearly three decades. He is a witness and explorer of the development of Chinese design art. As China's first generation of interior designers, he witnessed a take-off era, also witnessed the rise of interior industry.

In 2001, Mr. Dickson Hung named his own name to found the Dickson Hung Design Consultants (HK) Limited. Six years ago, he set up DHA International Design Alliance team to enhance the cooperation with the European designers, and made many foreign designers friends.

He takes a play attitude to design, and often says that he plays design nearly three decades, the first fifteen year, he was on the project site, and later, over ten years, he played design through exchanging design idea in Hong Kong, China, Europe and other areas. So far, Mr. Dickson Hung still on the way to explore different design types. We admire such an attitude to design, and we also see that design is a real pleasure and satisfaction for him.

With the 30-year of design art, Mr. Dickson Hung does design works by heart, and makes a lot of friends. In the minds of friends, he is enthusiastic, hard working, and his works are fantastic, and he is also a perfect partner.

（1）

Dickson Hung and I have known each other for over 15 years. At that time he gives me the impression that he is the person who works very hard; Secondly, he has an upward attitude to the interior design industry, he is very curious about design, wants to know the interior design situation around the world. He has this passion and keeps it all time. In addition, he is very sincere and friendly.

中国深圳国贸大厦——中国改革开放的标志性建筑，在中国室内设计行业发展史上也有着里程碑式的意义。从参与国贸大厦的室内装饰设计开始，洪德成（Dickson Hung）先生的室内设计生涯已走过近三十年。他是中国设计艺术发展的见证人，也是探索者。作为中国第一代室内设计师，他亲历了一个时代的腾飞，也见证了一个行业的崛起。

2001年，洪德成以他个人的名字创办了洪德成设计顾问（香港）有限公司。六年前，他又组建了DHA国际设计联盟团队，加强了与欧洲设计师之间的合作，也和不少国外设计师成了好朋友。

他常常说自己在"玩"设计，"玩"了近三十年，前十五年多半在工地上，后十多年，却是在与中国香港、欧洲等的交流气氛中"玩"设计。到目前为止，我们看到的洪德成先生还在探索不同的设计类型。这种心态令人敬佩，也可以由此看出，设计于他，真正就是一种乐趣和满足感。

三十年的设计艺术，用心做设计的洪德成先生结识了很多的朋友，在朋友们的心目中，他待人热情，工作勤力，他的设计不随波逐流且恰到好处。同时，他也是一个百分百好的合作伙伴。

（一）

我与洪德成认识已有15个年头。当时他给我的印象，一是工作很勤力；二是对这个行业有一颗向上的心，有关设计的他都会问，看外面是怎么样，他有这个热情，到现在也有这种热情；另外，他对人是诚恳、亲切的。

从中国改革开放初期到现在，洪先生从事这个行业这么多年，到他这个年纪，有钱、有名誉，很多人觉得他不需要再学习了，但是洪先生不同，他可以把项目放下，自己一个人到别的地方学。这个我认为不是每个人

Since Mr. Hung engaged in the interior design industry at the beginning of the Chinese reform and opening up, it has lasted for many years, he still insists in learning at his age, not for commercial benefits and reputation. He can leave the project alone to go on his study at another place. In my opinion, not everyone can do this, when we have related matters of family, friends and job, it's hard for me, so I am very envious of him.

Impressed by Willton Kwok, Design Director of Hong Kong Zhuoer Design Consultant Company

(II)

Time flies, it has been over ten years since we knew each other. At that time, I was a designer of Hong Kong architectural firm, I went to Shanghai to buy materials under the requirement of the customer, and I met Mr. Dickson Hung first time, he was the design director of another design company involved in this project. So we had an opportunity to work together for several days, we became good friends during those days.

In my mind, Mr. Dickson Hung is friend and teacher for me. He treats everyone friendly though he has achieved rich experience in design art industry. By the way, the passion for life and spirit of sharing make his infinite charm. We often discuss that Mr. Dickson Hung is a born artist and designer, the intellectual erudition due to his personality of interesting in everything, the design inspiration due to his passion for travel and photography. My favorite and most enjoyable moment is sharing travel photos with him, looking back the journey time in the busy time. For me, it is spirit supplements to ease work pressure and inspire new idea.

I appreciate that he never follows the crowd and does the right design, there are many designers who are following the trend to design, but his work is always filled with the theme of "people-oriented". His good communication skill and extensive intelligence are contributed to his unique projects, endowed custom-made masterpiece with soul. His design career and his

都可以做到的，到这个年纪，有家庭、有朋友、有工作——我也不可以，我很羡慕他这样做。

香港卓尔设计顾问公司设计总监 郭炳耀

（二）

转眼间我与洪德成先生已经结识十多个年头了。相识之初，我在香港某大建筑师楼担任设计师一职，当时客户要求我前往上海选购物料，谁知道这个项目亦有另一家设计公司参与，洪德成先生就是那家设计公司的设计总监，这样我和他一路共处了好几天，亦因此成为好朋友。

洪德成先生于我亦师亦友。在行内充满经验的他不但丝毫没有架子，热爱生活和分享的精神更使他充满无限魅力。他是天生的艺术家和设计师，因为他对事事感兴趣的性格，为他带来博学的知识，对旅游和摄影的热爱，成为了他设计的来源。我最喜欢和他分享旅游照片，这是我最享受的时刻，使我在繁忙中抽离，好好回味旅途中的片段。对我来说，这是很适合的精神补充剂，既可以缓和工作压力，又从中启发新思维。

我很欣赏他不随波逐流且恰到好处的设计，行内很多设计师只会应潮流而从事设计，但他的作品，往往充满"以人为本"的情结。他善于沟通，拥有广博的知识，促成他的项目都是独一无二、赋以灵魂的 custom made 杰作。他的设计事业已与他的生命融为一体，设计是他生命的一部分，因此，他的作品充满不同的情感，无论是哪一个文化背景的人，都能体会到他所表达的情感，这才是设计的最高境界。

刘伟婷设计师有限公司设计总监 刘伟婷

IMPRESSION OF DICKSON HUNG
朋友眼中的洪德成

life have been blended together, design is a part of his life, each work has different emotion. Regardless of the cultural background, the customer can always understand his expression — this is the highest level of design.

    Impressed by Flora Lau, the Design Director of Flora Lau Design Company Co. Ltd.

(III)

Mr. Dickson Hung treats us with sincerity and respect; he also is a far-sighted and ambitious person. His open-minded and passion for design is unquestionable. As a partner, Dickson is a perfect partner. He absolutely respects for others' design, never pretends to modify others' design work or impose his own intentions to others, to make mutual trust and respect. Because of the trust and respect, I always invite Dickson to give me some advice to get the better design. In my mind, Dickson is different from other designers I knew, he has an irresistible charm, he wants to step into the international stage, and no one can stop him.

    Impressed by NEVIN, the Design Director of A & N Architect & Interior Designer Limited

(IV)

I had know Dickson many years ago and I remember that day because I was in China for many years but, all Chinese Interior and Architect I'd seen before was very different from Dickson… I remember that day because I was been in Dickson Company and I had talked a lot with this interior design, his brother and his wife. I remember a good me with a good family and I understood that this man can be a good friend for me.

For the first time in China I can talk about design and furniture in the same mode as us did in Italy because Dickson understood what I think about and better Dickson had the same idea. Then during our meeting we was really good friends and every time that we discuss a work together we can have good idea and good job.

（三）

Dickson 是一个待人以诚，尊重别人，有远见而且有志气的人。他思想开放，对设计的热忱不可置疑。作为一个合作伙伴，Dickson 是个百分百的好对象。他对别人的设计绝对尊重，从不妄自修改他人的设计或硬要把自家的意图强加给别人，使大家互相信任及尊重。就是因为这份信任及尊重，我更会主动邀请 Dickson 提出意见，把设计做得更好。Dickson 跟我认识的其他设计师很不同，他有着一股挡不住的魅力，他要向世界冲出去，他想得到，他办得到。

    墺错设计有限公司何智诺设计总监 何智诺

（四）

我和洪先生已经认识多年了。我对洪先生的第一感觉是，他和中国其他设计师很不一样。我记得那次我前往洪先生的公司进行参观交流，我跟他和他的哥哥、夫人就室内设计进行了一次详细的交流。当时的我看到洪先生有着一个幸福美好的家庭，我觉得很高兴。也因为这样，我确信我日后一定可以和洪先生成为好朋友。

和洪先生第一次见面我是按照我们意大利设计师的思维方式和他进行交谈的。我很惊讶洪先生的思维方式和意大利设计师有着如此相似的地方，所以那次我们的交谈取得了很大的成果。随后，我们确实成为了好朋友并进行了实质性的项目合作。

我觉得洪先生和他的公司都具有开放性的特点，这个开放性和我们意大利人的思维方式是相吻合的。正是有着这个共同之处，我想我和洪先生在未来的合作上定能取得更大的成绩。 我认为洪先生本人在未来将会有更出色的成就。

    马可·丹尼卡洛（意大利著名室内设计师、家具设计师）

Now I can tell that Dickson and his Company is a very good Interior design with open mentality very near to mine Italian mentality. So I am very happy for this meeting and I think that we can build a lot in future together. Ten years are a lot but are nothing for all that this man can do…

Impressed by Marco Denicolò, the Famous Italian Furniture & Interior Designer

(V)

We knew each other around the year of 2001; there was no exact concept of decoration at that time. But Mr. Dickson Hung invited us to do decoration design service for a show flat project in Shenzhen with his far-reaching mind. Mr. Dickson Hung keeps his young and open mind but not the relatively conservative attitude as most Chinese people. He often exchanges and communicates with other designers from Hong Kong, China and European countries; he is a good example for us.

He is a designer, rather than a business man to look for commercial benefits. He has a very high demand for each project and often has lots of innovative ideas to meet the requirement of every work, never considers the commercial benefits. For designers, we satisfy with the efforts, the exchange with other designers. Mr. Dickson Hung is a good friend that worth making, not about any commercial benefit.

Mr. Dickson Hung has started hotel design. I do wish him a success in this area. And I believe that he has the ability to retain his insistence on the design, and make a harmonious balance.

Impressed by Teri Yeung, the Designer of Creative Arts Workshop limited Company

(From Zhonghua Interior Design Website)

（五）

我和洪总是2001年左右认识的，那时候国内还完全没有软装设计概念和行业，洪总却有非常前沿的眼光，请我们从香港到深圳来，帮他的一个样板房项目做软装配饰的摆场服务。多数中国人的思想相对会比较保守，但是洪总一直保持很年轻、很开放的心态，经常跟中国香港、欧洲的设计师交流，这是非常值得学习的一种精神。

他是一个真正的设计师，而不是一个追求商业利益的人，他对每个项目都有很高的要求，在我们的合作过程中，他经常会有很多创新的想法，永远是以达到设计效果为主，而不是看能赚多少钱。大家做设计的人，有付出，有交流，才有满足感。这也是我们这么多年都可以做朋友的一个原因，我们不是以商业的眼光去衡量，而是觉得这个朋友值得交。

洪总这几年开始做酒店了，我祝福他在这个领域里面可以做出很好的成绩，同时还可以保留自己对设计的坚持，要平衡这两者很不容易，我相信他有这个能力。

创艺工作坊有限公司设计师 杨洁仪

（摘自中华室内设计网）

Maternal Love in the Beginning of Art

母爱，在艺术的起点

In those days, mother's embrace always gave us warming feeling. In my memory, mother was very diligent, went through ups and downs to bring us up. When I was only six years old, whenever I heard the sound of footsteps on the stairs, I certainly could distinguish hers and she always gave me something to eat. Mother was born in a family of intellectuals, with good genes and educational background. Since she had been married with my father, she resigned from her excellent work in order to raise us five children, and dedicated to training us. Mother is an esthetician, her feeling and demands to beauty are natural. My elder brother is a painter, and I became a design artist, all these should thanks to the "moral" education of mother. From childhood until this Spring Festival, I would like to put my favorite paintings and design work and show them to mother, she always said, "beautiful." My mother, 90-year-old, her ending was peaceful. There is no chance for me to communicate with mother face to face and shake hands with her. In my opinion, mother was very great, she went through different hard times with courageous, prudent, positive and optimistic attitude to face everything; her spirit and soul always go with our family.

那个年代，在母亲的怀抱里，是那样的温暖！我自懂事起，就记得母亲非常勤快，风风雨雨，就为了让我们兄弟姐妹长大成人。那时我才6岁，只要听到上楼梯的脚步声，我就一定听得出是母亲，而且母亲每次都带着吃的给我。母亲出生在一个知识分子家庭，有着很好的基因和教育背景，但自从和我父亲成家后，为了抚养我们5个子女，辞去了优越的工作，全心培养我们。母亲是个美学家，她对美的感觉和要求都是天生的。我的大哥是个画家，而我也是个设计艺术家，这都归功于母亲的"德行"教育。从小时候开始一直到今年春节前，我最喜欢将我的绘画作品和设计作品拿给我母亲看，她总是说"漂亮"。母亲90岁高寿，早前在家中去了。我不能再面对面和母亲握着手交流了，但母亲的一生是伟大的，她走过了千山万水的艰辛，有勇有谋、积极乐观地面对一切，她的精神、她的灵魂永远和我们家人在一起。

To My Parents 谨以此书献给我的父母亲

17 year-old mother and father 17岁的母亲和父亲

I was 4 years old 4岁的我

On the left, the first one is my great-grandfather, mother is standing on the bench, on the left, the third one is grandfather; the early 1940s, parents went back to the mainland from Hong Kong. 左一是我的太老外公，站在长椅上的是母亲，左三是外公。20世纪40年代初，父母从香港回到内地。

ents and my four brothers and sisters, I am their fifth child, I was not born at that time. 父母与我的四个哥哥姐姐，我排行老五，此时还未出生。